Portville Free Library
Portville, New York 14770

DISCARDED FROM THE
PORTVILLE FREE LIBRARY

S0-EIR-018

Ice-Age Monsters
Saber Tooth Tiger

Written by Rupert Oliver
Illustrated by Roger Payne

© 1986 Rourke Enterprises, Inc.

All rights reserved. No part of this book may be reproduced or utilized in any form or by any means, electronic or mechanical including photocopying, recording or by any information storage and retrieval system without permission in writing from the publisher.

Library of Congress Cataloging in Publication Data

Oliver, Rupert.
  Saber Tooth Tiger.

  Summary: Describes a day in the life of a prehistoric Saber Tooth tiger as he searches for water and food and tries to avoid danger.
  1. Smilodon—Juvenile literature. [1. Saber-toothed tigers] I. Payne, Roger, fl. 1969-   ill. II. Title.
QE882.C15044 1986   569'.74   86-557
ISBN 0-86592-845-2

Rourke Enterprises, Inc.
Vero Beach, FL 32964

Ice-Age Monsters
Saber Tooth Tiger

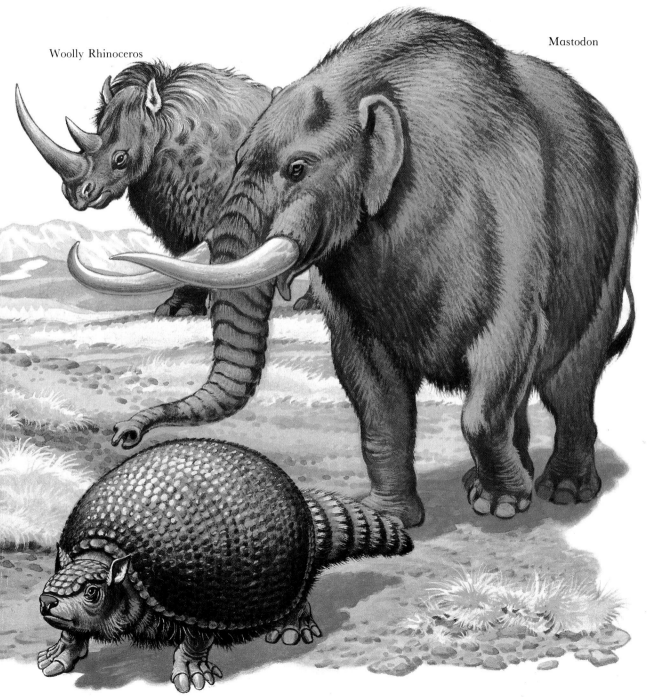

The bushes rustled as a small bird leaped from branch to branch. It was Spring and the bird had to find a mate to help build a nest. Suddenly, the bird burst forth into song.

The chirping bird was making so much noise that he woke up a great, furry animal which had been sleeping under the bushes. As the large animal stirred and yawned, the bird stopped singing and watched.

The powerful Saber Tooth Tiger rose and smelled the air. He could smell blood and it was not very far away. He lifted his head and roared in satisfaction and the bird flew off in alarm. Saber Tooth moved off through the bushes, following the smell of blood.

As Saber Tooth emerged from the bushes he saw a strange sight. Just below him was a great pool of liquid, and in the pool was a group of snarling, fighting animals. In the middle of the pool was a great shaggy bison and all around it were hungry wolves.

The wolves were trying to kill the bison and had already wounded it several times. It was the blood from these wounds which Saber Tooth had smelled. Even though the bison had been injured, it was still fighting and was able to keep the wolves at bay. Saber Tooth remembered how good the taste of bison was and roared again. Then, he moved forward to join the wolves in the fight.

As Saber Tooth approached the pool of liquid he noticed something was wrong. The liquid was not splashing around like water. With all the fighting going on, there should be much spray. Then he noticed something else. All the animals in the pool were becoming stuck in a thick black liquid. Already the bison was beginning to sink into the pool and obviously could not get out.

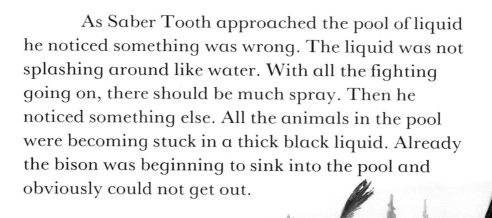

Perhaps if Saber Tooth joined in the fight he would also become stuck. Saber Tooth decided to leave the bison and wolves alone. He strolled off across the plain to look for something to drink. As he left, a vulture swooped down out of the sky. It had come to take any meat which the wolves might leave behind. Instead it, too, became trapped in the pool.

As Saber Tooth walked through the grass, he noticed a sudden flurry of movement. To his right was a herd of Pronghorns. The Pronghorns had been watching Saber Tooth in case he decided to attack them. They had not seen a Felis Trumani creeping up from the other side.

Suddenly, the Felis Trumani leaped from cover and began to chase the Pronghorns. Soon, it had singled out one Pronghorn which was slower than the rest. The Pronghorn bounded across the plain as quickly as it could, but Felis Trumani could run very fast indeed.

As the two animals raced past Saber Tooth, the Felis Trumani was catching up with the Pronghorn. Then, the Pronghorn put on a extra burst of speed and the Felis Trumani gave up.

Saber Tooth watched the Felis Trumani standing exhausted. Saber Tooth could not run as fast as Felis Trumani and had never been able to catch a Pronghorn. He did not even bother trying to attack the Pronghorns, instead he moved off toward the river. Saber Tooth was thirsty so he hurried toward the water. He followed a path down to the river.

The path had fresh hoof marks, but Saber Tooth was more interested in water than food, so he did not attempt to move quickly. As he approached the water, a group of horses saw Saber Tooth. They neighed in alarm and galloped across the river to reach safety. Saber Tooth walked down to the river and drank his fill.

When Saber Tooth had finished drinking he realized that he had not had anything to eat since the previous day. He was very hungry. He moved away from the river and toward the grasslands. As Saber Tooth came over a rise in the land he saw a pile of meat lying on the ground. There was more than enough meat there for Saber Tooth. He walked toward the meat.

As he approached the meat, Saber Tooth suddenly halted in his tracks. Resting close to the meat were two of the only animals Saber Tooth feared. They were men. They were armed with spears which Saber Tooth knew to be dangerous. When the men saw Saber Tooth they started shouting and waving their spears. Saber Tooth snarled at them, then turning, he ran away. Even the savage Saber Tooth was frightened of man.

Saber Tooth was still hungry. He would have to go in search of living prey. Saber Tooth moved back toward the river. It was a very hot day and many animals would come down to the water to drink. Perhaps Saber Tooth would be able to catch one of them.

Saber Tooth moved down the path to the river until he came to a thick clump of bushes. Then, he moved off the path and hid himself in the bushes. From the bushes he would be able to spring out on anything which went past.

Saber Tooth settled down. He did not have to wait long. A group of horses moved down the path, but they smelled Saber Tooth and sped away for dear life.

It was a long time before another animal took the path toward the river. Then, Saber Tooth heard heavy footsteps. Something large was coming to drink. It did not run away so it could not have smelled Saber Tooth. Saber Tooth prepared to spring on the unwary newcomer.

Then, Saber Tooth saw what was coming. It was a huge Megatherium. As soon as the animal was close enough, Saber Tooth sprang out with a ferocious roar. The Megatherium heard the roar but before it could do anything to protect itself, the massive teeth of the huge cat were plunging deep into its back.

When Megatherium was dead, Saber Tooth began to eat. Saber Tooth ate as much of the fresh succulent meat as he could manage. Then, he walked off into the bushes to sleep.

Meanwhile, a pack of wolves moved in to finish picking the meat off the Megatherium. When Saber Tooth woke up there would be little left of his kill.

Saber Tooth Cats and their times

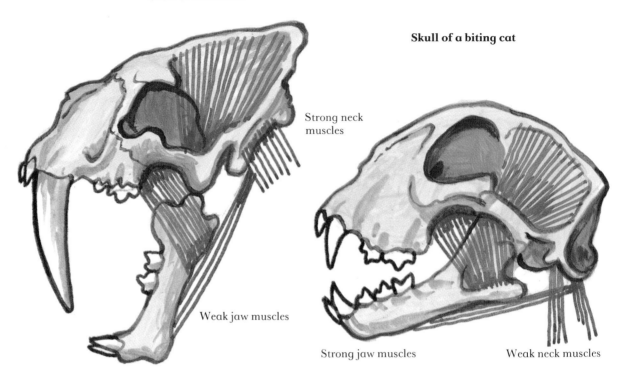

Skull of Smilodon

Strong neck muscles

Weak jaw muscles

Skull of a biting cat

Strong jaw muscles

Weak neck muscles

When did the Saber Tooth live?
There were many different species of Saber Tooth cats which lived at many different times. The earliest Saber Tooth cats lived about 30 million years ago and they lived all over the world, except in Australia. The Saber Tooth cat in the story is known today as Smilodon. Smilodon lived during the Pleistocene era, that is during the past two million years. There are no Saber Tooth cats alive today since they all died out about 11,000 years ago.

The Saber Tooth
Cats first appeared about 35 million years ago and almost from the start they divided into two lines. One was that of the biting cats. All cats alive today, including lions and tigers, belong to this group. The other group were the stabbing cats. Smilodon and all Saber Tooth cats were stabbing cats. Stabbing cats could not run as fast as biting cats and so hunted slower prey. In the story you can see the Felix Trumani running faster than Smilodon. In order to hunt large, slow-moving prey, the stabbing cats evolved their massive canine teeth which have given them the name of Saber Tooth cats. Biting cats kill their prey by biting, but large prey cannot be killed so easily. The Smilodon therefore used its saber teeth to rip into prey. It is probable that Smilodon would plunge its fearsome teeth into its prey's neck and then use its powerful neck muscles to sever arteries and other blood vessels. Smilodon would then release its death-grip and wait for its prey to die from loss of blood. Since Smilodon died out some 11,000 years ago no cat has had such large teeth nor such powerful neck muscles.

Where did Smilodon live?
Smilodon lived in many parts of the world and was a very successful hunter. It even manged to reach South America, a land never reached by any cat before. The Smilodon in our story lived in North America, in what is now known as the state of California.

What were the tar pits?

The strange pools of liquid which the Smilodon came across early in our story were actually tar pits. Tar pits occur when oil seeps up from deep underground and collects in hollows on the surface. Any rain which falls onto a tar pit collects in pools on the tar's surface. Any animal passing which is thirsty is more than likely to attempt to drink this water. As soon as it steps onto the tar it is doomed. The weight of the animal drags it down into the sticky tar and it is unable to escape.

The tar pits of California turned to solid tar many years ago. In the tar were preserved the bones of all the animals which became trapped in the tar. Scientists have now dug up the tar and recovered the bones from the tar. Tar pits are therefore very important to scientists who study extinct animals.

There is one curious fact about the bones in the tar pits, and that is that there are far more meat-eaters, carnivors, than plant-eaters, herbivores. In reality there were far more plant-eaters than meat-eaters in California at the time. This anomaly can be explained by looking at what may have happened when an animal became trapped. A large animal such as a bison or mammoth would soon attract carnivores such as wolves or large cats. Any carnivore which stepped into the tar to attack the trapped animal would become trapped itself. It is for this reason that so many meat eaters have been found in the tar.

Other animals

Our story is set some 30,000 years ago. Though this is a long time to humans, it is only a short period in the history of the world. Many of the animals in our story can be seen today. The bison and Pronghorn are still to be found, as are the snakes, insects and birds of the time. Other animals have died out and are strange to us. One of these is the Megatherium. This beast was related to the tree sloths which survive today. In fact, it is often referred to as the giant ground sloth. It moved slowly and awkwardly and ate leaves from the trees. Some Megatheriums may have been as much as 18 feet long. Dire wolves have also become extinct, as has the Imperial Mammoth, an animal most people do not expect to see in America. The Imperial Mammoth, grew to be about 12 feet tall and had huge curving tusks which could measure a staggering 13 feet in length. Though the elephant is not usually associated with America, it is thought to have survived in South America until just a thousand years ago.

Dinictis. This ancestor of both stabbing and biting cats lived about 35 million years ago.

4-6

J 569.74 O
Oliver, Rupert.
    Saber Tooth tiger

DATE DUE		
DISCARDED FROM THE		
PORTVILLE FREE LIBRARY		

PORTVILLE FREE LIBRARY

PORTVILLE, N. Y.

Member Of
Chautauqua-Cattaraugus Library System